CARBON ZERO

HARNESSING TECHNOLOGY FOR EFFECTIVE CARBON CAPTURE

DIPAN KUMAR DAS

SUDIP KUMAR DAS

Dedication: To the Stewards of Our Planet's Future

This book is dedicated to all those who hold the well-being of our planet close to their hearts. To the scientists and researchers who tirelessly pursue breakthroughs in carbon capture technologies, your dedication paves the way for a more sustainable future. To the policymakers who craft forward-thinking regulations and incentives, your efforts drive the adoption of these vital solutions. To the innovators and entrepreneurs who dare to dream of a carbon-zero world, your vision sparks transformative change.

To the communities and individuals who make conscious choices every day to reduce their carbon footprint, your actions ripple out, creating a wave of positive impact. To the educators and communicators who bridge the gap between science and the public, your work empowers informed decisions and fosters a collective commitment to our planet.

To the leaders who forge international collaborations and partnerships, your unity amplifies our capacity to address global challenges. To the generations past, present, and future, your shared responsibility connects us through time as caretakers of this remarkable planet.

May this dedication be a tribute to your dedication—to the stewards of our planet's future who recognize that together, we hold the power to shape a world where sustainable practices and carbon capture technologies transform our challenges into opportunities. Your tireless efforts inspire us all to work toward a brighter, more resilient, and carbon-zero future.

Foreword

Foreword: Navigating the Path to Carbon-Zero Horizons

In a world grappling with the urgent need to address climate change, the pursuit of carbon-zero solutions has emerged as a beacon of hope. As we stand on the precipice of transformative change, "Carbon Zero: Harnessing Technology for Effective Carbon Capture" guides us through the complexities and opportunities that lie ahead.

This book serves as a roadmap, offering insights into the scientific foundations of carbon capture technologies, the innovative strategies for their implementation, and the societal dynamics that shape their adoption. It brings together knowledge from diverse disciplines, uniting the realms of science, policy, economics, and human behavior.

As we venture deeper into the chapters, we are reminded that the journey to carbon-zero is a collective endeavor. The successes of countries, cities, organizations, and individuals inspire us to transcend boundaries and seek collaborative solutions. The case studies demonstrate the transformative power of technology, showing how innovation can reshape industries, cities, and the global energy landscape.

Equally important is the discussion of public perception and education. The book underscores the role of communication in fostering understanding, acceptance, and support for carbon capture technologies. It reminds us that the path to a sustainable future is not solely paved with technological breakthroughs but is also illuminated by informed choices and a shared commitment to change.

As you embark on this journey through the pages of "Carbon Zero," I invite you to reflect on the possibilities it presents. Consider the role you can play in accelerating the transition to carbon-zero horizons—whether as a scientist, a policymaker, an entrepreneur, an educator, or an engaged citizen. The future is in our hands, and it is the collective action of individuals that will define our success.

With a united vision, a thirst for knowledge, and a resolute determination, we can navigate the complexities of carbon capture technologies and steer humanity towards a future where carbon emissions are harnessed, and our planet thrives in balance.

Preface

Embarking on a Journey to Carbon-Zero Solutions

In an era defined by unprecedented environmental challenges, the need for effective solutions to combat climate change has never been more urgent. As we grapple with the complex interplay of science, technology, policy, and societal dynamics, "Carbon Zero: Harnessing Technology for Effective Carbon Capture" embarks on a journey of exploration, discovery, and enlightenment.

This book is a culmination of extensive research, collaboration, and a shared commitment to addressing one of the greatest challenges of our time. It delves into the world of carbon capture technologies, shedding light on their potential to revolutionize energy systems, industries, and our very way of life. Each chapter unfolds a new layer of understanding, taking readers from the scientific foundations of carbon capture to the innovative applications that hold the promise of a more sustainable future.

As you navigate through these pages, you will encounter a rich tapestry of insights, case studies, and perspectives that illuminate the multifaceted nature of carbon capture. From the intricacies of various capture methods to the implications of widespread adoption, this book provides a comprehensive lens through which to view the path towards carbon-zero horizons.

But this journey is not confined to the realms of academia and technology. It extends to our communities, our choices, and our aspirations for the future. It is a journey that invites us to reflect on our roles as stewards of the planet, to engage in conversations that bridge knowledge gaps, and to inspire action that brings about meaningful change.

As the authors of this book, we are humbled to be your guides on this journey. We invite you to immerse yourself in the narratives, to contemplate the possibilities, and to envision a world where carbon capture technologies are not just solutions, but agents of transformation. Together, we have the power to shape a future that is not burdened by the weight of carbon emissions, but is fueled by innovation, collaboration, and a shared commitment to a sustainable world.

Welcome to the journey of "Carbon Zero: Harnessing Technology for Effective Carbon Capture." May the insights you gain and the stories you encounter inspire you to become a part of the movement towards a carbon-zero future.

Prologue

The Imperative for Carbon-Zero Solutions

In the backdrop of a changing climate and a world grappling with the consequences of carbon emissions, the need for transformative solutions has never been more pressing. As we stand at this pivotal juncture in history, the prologue of "Carbon Zero: Harnessing Technology for Effective

Carbon Capture" invites us to contemplate the urgent imperative of addressing carbon emissions and the potential of carbon capture technologies to reshape our trajectory.

The prologue sets the stage by outlining the far-reaching implications of unchecked carbon emissions. It highlights the complex web of challenges that extend beyond environmental impact, affecting economies, societies, and ecosystems. It underscores the importance of recognizing carbon capture as a critical tool in mitigating climate change, reminding us that a sustainable future hinges on harnessing human ingenuity to address this global challenge.

As you embark on this journey through the chapters that follow, remember that this is not merely a scientific exploration; it is a call to action. It is an invitation to envision a world where carbon emissions are no longer a liability but a resource—a resource that can be captured, transformed, and harnessed for the betterment of our planet and future generations.

With a renewed sense of purpose, let us venture into the narratives that lie ahead. Let us seek to understand the intricate mechanisms of carbon capture, the strategies for its effective implementation, and the collective efforts required to usher in a carbon-zero era. Together, we can rise to the challenge and work towards a future defined not by carbon emissions, but by the innovative solutions that pave the way to a more sustainable and resilient world.

As we turn the page to "Carbon Zero," let us remember that the journey to a carbon-zero future begins with each of us—an individual, an advocate, a catalyst for change. Let the prologue inspire you to embark on this journey with an open heart and an unwavering commitment to making a difference.

CHAPTER ONE

Understanding Carbon Emissions and Their Impact

Introduction

In an era defined by environmental awareness and the pressing need to address climate change, understanding carbon emissions and their far-reaching impact is paramount. This chapter serves as a foundational exploration of carbon emissions, elucidating their role in driving climate change, their sources across various sectors, and the consequences of unchecked emission levels.

1.1 The Carbon Connection to Climate Change

The global climate is intricately connected to the balance of greenhouse gases in the atmosphere, with carbon dioxide (CO_2) playing a pivotal role. This section introduces readers to the greenhouse effect and explains how certain gases trap heat, leading to a rise in global temperatures. It highlights CO_2 as a significant contributor to this effect and outlines the scientific consensus linking increased CO_2 levels to the observed warming trends.

1.2 Sources of Carbon Emissions

Diverse sectors contribute to carbon emissions, ranging from energy production and transportation to agriculture and industrial processes. This section provides a comprehensive overview of these sectors and discusses their respective roles in releasing CO_2 and other greenhouse gases into the atmosphere. The sources of emissions are discussed in terms of both direct emissions, such as from tailpipes and smokestacks, and indirect emissions, including deforestation and land use changes.

1.3 Impact on the Environment, Economy, and Society

The consequences of unchecked carbon emissions reverberate through multiple aspects of human life. This section delves into the multifaceted impacts of escalating carbon emissions, including:

Environmental Changes: Discusses the intensification of extreme weather events, sea level rise, and disruptions to ecosystems due to changing climatic patterns caused by increased carbon emissions.

Economic Implications: Explores the economic costs associated with climate change, including damage to infrastructure, loss of productivity in agriculture, and the financial burden of adapting to changing conditions.

Social Challenges: Addresses the disproportionate effects of climate change on vulnerable populations, including communities in low-lying coastal areas, developing nations, and marginalized groups.

1.4 The Call for Action: Transitioning to Carbon Neutrality

With the urgent need to mitigate climate change, transitioning to a carbon-neutral society is imperative. This section emphasizes the importance of reducing carbon emissions and provides a glimpse into the concept of carbon neutrality, wherein a balance is achieved between carbon emissions and carbon removal from the atmosphere. It introduces the concept of carbon capture and storage (CCS) as one of the key strategies to achieve this balance.

Conclusion

Chapter 1 establishes the foundation for understanding the relationship between carbon emissions and climate change. By exploring the fundamental concepts of greenhouse gases, emission sources, and the consequences of unchecked emissions, readers are primed to delve deeper into the subsequent chapters, which will focus on the technologies and strategies aimed at mitigating the impacts of carbon emissions and fostering a sustainable future.

1.2 Overview of the Sources and Sectors Contributing to Carbon Emissions

Carbon emissions originate from a diverse array of sources spanning various sectors of human activity. Understanding these sources is essential for devising effective strategies to reduce emissions and mitigate their impact on the environment. This section provides an in-depth overview of the primary sources and sectors that contribute to carbon emissions.

2.1 Energy Production and Consumption

One of the most significant contributors to carbon emissions is the burning of fossil fuels for energy production and consumption. This includes the combustion of coal, oil, and natural gas to generate electricity, power vehicles, and fuel industrial processes. Power plants, factories, and transportation systems emit substantial amounts of carbon dioxide and other greenhouse gases during fuel combustion.

2.2 Transportation

The transportation sector encompasses road, air, rail, and maritime travel. Combustion engines in cars, trucks, airplanes, ships, and trains release carbon emissions into the atmosphere. This section explores the impact of increased urbanization, global trade, and rising demand for mobility on transportation-related emissions and discusses potential solutions, including electric vehicles, public transportation, and improved fuel efficiency.

2.3 Industrial Processes

Industrial activities, such as manufacturing, chemical production, and cement manufacturing, release carbon emissions both directly and indirectly. Indirect emissions result from the energy-intensive processes involved in creating products and materials. This section discusses the role of industrial emissions in the overall carbon footprint and examines technological advancements and innovations that aim to reduce emissions from these sectors.

2.4 Agriculture and Land Use Changes

Agriculture contributes to carbon emissions through multiple avenues. Livestock farming generates methane emissions from animal digestion and manure, while rice paddies emit methane due to anaerobic conditions. Additionally, deforestation and land use changes release stored carbon into the atmosphere. This section explores the complex relationship between agriculture, land use, and emissions, highlighting practices like sustainable agriculture, afforestation, and reforestation.

2.5 Residential and Commercial Buildings

Energy consumption in residential and commercial buildings for heating, cooling, lighting, and appliances leads to carbon emissions. The use of fossil fuels for heating and electricity, along with inadequate insulation and outdated infrastructure, can significantly contribute to emissions. This section discusses energy-efficient building designs, renewable energy integration, and behavioral changes that can collectively reduce emissions from the built environment.

2.6 Waste Management

Waste management processes, including landfilling and incineration, release methane and carbon dioxide as organic waste decomposes. The improper management of waste exacerbates emissions. This section examines strategies such as waste reduction, recycling, composting, and advanced waste-to-energy technologies that can minimize emissions from waste disposal.

2.7 Deforestation and Land Degradation

Forests act as carbon sinks by absorbing carbon dioxide from the atmosphere. However, deforestation and land degradation release stored carbon, contributing to emissions. This section highlights the importance of preserving and restoring forests, as well as adopting sustainable land management practices to mitigate emissions associated with land use changes.

Conclusion

Chapter 1.2 sheds light on the diverse sources and sectors that collectively contribute to global carbon emissions. Recognizing the interconnectedness of these sources is crucial for implementing targeted strategies to mitigate emissions and transition toward a more sustainable and carbon-neutral future. By addressing emissions across sectors, it becomes possible to develop comprehensive solutions that tackle the root causes of carbon emissions and their impact on the environment.

1.3 Discussion of the Environmental, Economic, and Societal Consequences of Unchecked Carbon Emissions

The unmitigated release of carbon emissions into the atmosphere has far-reaching implications that extend beyond environmental concerns. This section delves into the interconnected consequences of unchecked carbon emissions, including their impact on the environment, economy, and society.

3.1 Environmental Consequences

Climate Change and Global Warming: The foremost consequence of unchecked carbon emissions is climate change, resulting in rising global temperatures, altered weather patterns, and increased frequency and intensity of extreme weather events. These changes disrupt ecosystems, threaten biodiversity, and jeopardize the delicate balance of nature.

Sea Level Rise: As global temperatures rise, polar ice melts and thermal expansion of seawater occurs. These factors contribute to rising sea levels, which endanger coastal communities, exacerbate flooding, and erode coastlines. Low-lying regions are particularly vulnerable.

Ocean Acidification: Carbon emissions not only warm the atmosphere but also dissolve into oceans, leading to ocean acidification. This process threatens marine life, including coral reefs and shellfish, by affecting the availability of calcium carbonate, a crucial building block for shells and skeletons.

3.2 Economic Implications

Infrastructure Damage: Increasingly severe weather events, such as hurricanes and flooding, can damage critical infrastructure like roads, bridges, and power grids. The resulting repair costs strain public budgets and hinder economic development.

Agricultural Disruptions: Altered weather patterns and extreme events can disrupt agricultural productivity.

Crop failures, reduced yields, and shifts in suitable growing regions can lead to food scarcity, price hikes, and economic instability.

Energy Costs and Reliability: Unchecked carbon emissions contribute to energy demand and supply challenges. Higher temperatures increase the need for cooling, while extreme events can disrupt energy production and distribution, leading to increased energy costs.

3.3 Societal Challenges

Health Impacts: Climate change fosters conditions conducive to the spread of diseases such as malaria, dengue fever, and heat-related illnesses. Vulnerable populations, including the elderly and low-income communities, are particularly at risk.

Migration and Displacement: Environmental disruptions caused by climate change can trigger forced migration and displacement of communities, leading to social tensions, conflicts, and strained resources in receiving areas.

Social Inequities: Unchecked carbon emissions exacerbate existing social inequalities. Marginalized and disadvantaged communities often bear the brunt of environmental impacts and have limited resources to adapt or recover.

3.4 Urgent Need for Mitigation and Adaptation

The multifaceted consequences of unchecked carbon emissions underscore the pressing need for mitigation and adaptation strategies. Transitioning to renewable energy sources, implementing carbon capture and storage technologies, and adopting sustainable land use practices are critical steps toward mitigating these consequences. Additionally, proactive planning and investment in resilient infrastructure, disaster preparedness, and social safety nets can enhance societies' ability to adapt to the changing climate.

Conclusion

Chapter 1.3 illuminates the intricate web of consequences resulting from the unchecked release of carbon emissions. These consequences touch every facet of our world, from the environment and economy to society's well-being. By comprehending the potential impacts of our actions, we empower ourselves to make informed decisions that safeguard our planet and its inhabitants, while fostering a sustainable future for generations to come.

1.4 Importance of Transitioning to Carbon-Neutral or Carbon-Negative Practices

The imperative to combat climate change and its associated

consequences has given rise to a critical need for transitioning from carbon-intensive practices to those that are carbon-neutral or even carbon-negative. This section highlights the significance of such a transition and the benefits it offers to the environment, economy, and global sustainability.

4.1 Mitigating Climate Change

Reducing Greenhouse Gas Emissions: Transitioning to carbon-neutral or carbon-negative practices involves minimizing the release of greenhouse gases, primarily carbon dioxide, into the atmosphere. This mitigation is essential to curbing global temperature rise and avoiding the worst impacts of climate change, including extreme weather events and sea level rise.

Preserving Ecosystems: Carbon-neutral practices promote the protection and restoration of forests, wetlands, and other natural ecosystems. These environments act as carbon sinks, absorbing atmospheric carbon and contributing to the overall balance of greenhouse gas concentrations.

4.2 Fostering Innovation and Technological Advancement

Incentivizing Research and Development: The transition to carbon-neutral or carbon-negative practices drives innovation in renewable energy technologies, carbon capture and storage methods, and sustainable land use strategies. These innovations not only mitigate emissions but also propel economic growth and competitiveness.

Creating Green Jobs: The shift towards carbon-neutral practices creates opportunities for employment in emerging sectors such as renewable energy, clean transportation, and ecosystem restoration. This contributes to local economies while simultaneously addressing environmental concerns.

4.3 Enhancing Energy Security and Resilience

Diversifying Energy Sources: Relying on renewable energy sources like solar, wind, and hydroelectric power reduces dependence on fossil fuels and enhances energy security. These sources are inherently less susceptible to supply disruptions and price volatility.

Building Resilience to Climate Impacts: Transitioning to carbon-neutral practices can bolster communities' resilience against the impacts of climate change. By investing in resilient infrastructure and disaster preparedness, societies are better equipped to withstand extreme events and recover more swiftly.

4.4 Global Cooperation and Diplomacy

Meeting International Commitments: The transition to carbon-neutral or carbon-negative practices is aligned with international agreements such as the Paris Agreement. By fulfilling commitments to emission reduction, nations foster global cooperation and contribute to collective efforts to limit global warming.

Mitigating Climate-Induced Conflicts: Addressing climate change through carbon-neutral practices can mitigate the potential for resource scarcity and environmental degradation-induced conflicts, contributing to global stability and peace.

4.5 Securing a Sustainable Future

Protecting Future Generations: Transitioning to carbon-neutral or carbon-negative practices is an ethical imperative to safeguard the planet for future generations. This proactive approach ensures a viable and habitable environment for all life forms.

Promoting Environmental Stewardship: Embracing carbon-neutral practices reflects a commitment to responsible environmental stewardship. It sets a precedent for responsible consumption and production patterns that respect the planet's finite resources.

Conclusion

Chapter 1.4 underscores the crucial importance of transitioning to carbon-neutral or carbon-negative practices as an essential step in addressing the challenges posed by climate change. By recognizing the environmental, economic, and social benefits of this transition, societies can work collectively to reduce their carbon footprint, mitigate the impacts of global warming, and pave the way for a sustainable and resilient future.

CHAPTER TWO

Fundamentals of Carbon Capture Technologies

2.1 Explanation of Different Carbon Capture Techniques

Carbon capture technologies are pivotal in the fight against climate change, as they offer a means to capture and store carbon dioxide emissions from various sources. This chapter provides a comprehensive understanding of the three primary carbon capture techniques: post-combustion, pre-combustion, and direct air capture.

2.1.1 Post-Combustion Carbon Capture

Post-combustion carbon capture is a technique used to capture carbon dioxide from the exhaust gases of power plants and industrial facilities after the combustion process. The core steps of post-combustion capture involve:

Gas Separation: Exhaust gases are cooled and compressed to separate carbon dioxide from other gases, such as nitrogen and oxygen, using absorption or adsorption processes.

Solvent Regeneration: The captured carbon dioxide is then released from the solvent using heat or pressure, allowing the solvent to be reused in subsequent capture cycles.

This technique is advantageous as it can be retrofitted onto existing power plants, making it a viable option for emissions reduction without major infrastructure changes.

2.1.2   Pre-Combustion Carbon Capture

Pre-combustion carbon capture is primarily employed in integrated gasification combined cycle (IGCC) power plants and facilities that convert solid or liquid carbon-based fuels into synthetic gas (syngas). The process involves:

Gasification: Carbon-containing fuels are converted into syngas through high-temperature reactions with steam or oxygen. This syngas comprises carbon monoxide (CO) and hydrogen (H2).

Gas Shift Reaction: The CO in the syngas is reacted with steam to form carbon dioxide and more hydrogen, increasing the concentration of carbon dioxide in the gas stream.

Carbon Capture: The concentrated carbon dioxide can be captured using various separation methods, such as absorption or membrane technologies, before combustion occurs.

Pre-combustion carbon capture is efficient because it allows for high-purity carbon dioxide streams and potentially lower energy penalties compared to post-combustion capture.

2.1.3 Direct Air Capture

Direct air capture (DAC) is a technique that involves capturing carbon dioxide directly from ambient air, rather than from an emission source. DAC systems utilize sorbents or solvents to chemically bind with carbon dioxide molecules. The captured carbon dioxide is then released, concentrated, and stored for utilization or geological storage.

DAC has the potential to remove historical carbon emissions from the atmosphere, making it a promising approach for achieving carbon-neutral or carbon-negative goals.

Conclusion

Chapter 2.1 provides a comprehensive overview of the three fundamental carbon capture techniques: post-combustion, pre-combustion, and direct air capture.

Each technique has its unique advantages and challenges, and understanding their mechanics is crucial for designing effective carbon capture systems. These techniques form the cornerstone of carbon capture and storage efforts, driving innovation and research to combat climate change and create a sustainable future.

2.2 In-depth Exploration of the Science and Engineering Behind Each Capture Method

Understanding the intricate science and engineering principles behind carbon capture methods is essential for developing efficient and effective technologies. This section delves deeper into the scientific and engineering aspects of each carbon capture technique: post-combustion, pre-combustion, and direct air capture.

2.2.1   Post-Combustion   Carbon Capture

Gas Separation: Post-combustion capture involves separating carbon dioxide from the flue gases emitted by power plants or industrial processes. This is achieved through absorption or adsorption processes. Absorption uses a liquid solvent that absorbs carbon dioxide from the flue gas, while adsorption employs solid adsorbents to capture the gas. The science behind these processes lies in the chemical interactions between the solvent or adsorbent and the carbon dioxide molecules.

Solvent Regeneration: After absorbing carbon dioxide, the solvent must be regenerated for reuse. This involves the application of heat or pressure to release the captured carbon dioxide. Engineering considerations include optimizing the temperature and pressure conditions for efficient regeneration without degrading the solvent's performance.

2.2.2 Pre-Combustion Carbon Capture

Gasification and Syngas Formation: In pre-combustion capture, gasification is a thermochemical process that converts carbon-containing fuels into syngas (CO and H2). This process involves high temperatures and controlled reactions. Engineering challenges include designing gasifiers that achieve high conversion rates and produce syngas with the desired composition.

Gas Shift Reaction: The gas shift reaction converts CO in the syngas to CO_2 and additional hydrogen. Catalysts are used to facilitate this reaction, and their selection and design impact the efficiency of carbon capture. The science involves understanding reaction kinetics and optimizing catalyst materials.

Carbon Capture: Separation techniques such as absorption or

membrane processes are employed to capture the carbon dioxide from the shifted syngas. The choice of separation method depends on factors such as energy efficiency and purity of the captured gas.

2.2.3 Direct Air Capture

Sorbents and Solvents: In direct air capture, sorbents or solvents are used to chemically bind with carbon dioxide from ambient air. The science involves designing materials with high affinity for carbon dioxide molecules while minimizing interactions with other components of the air. Engineering challenges include developing sorbents that are cost-effective and can be easily regenerated for reuse.

Carbon Dioxide Release and Concentration: Once the sorbent or solvent captures carbon dioxide, it needs to be released and concentrated for storage or utilization. This involves manipulating temperature and pressure conditions to facilitate desorption. Efficient release mechanisms are crucial for the viability of direct air capture systems.

Conclusion

Chapter 2.2 provides an in-depth exploration of the science and engineering principles underlying each carbon capture method. The intricacies of chemical reactions, materials design, thermodynamics, and process optimization are central to developing practical and scalable carbon capture technologies. By grasping these scientific and engineering fundamentals, researchers and engineers can innovate and refine carbon capture methods to effectively address the challenge of carbon emissions and climate change.

2.3 Case Studies Highlighting Successful Implementation of Carbon Capture Technologies

Real-world examples of successful carbon capture technology implementation demonstrate the feasibility and effectiveness of these approaches. This section presents case studies that showcase instances where carbon capture technologies have been deployed and their impact on emissions reduction and environmental sustainability.

2.3.1 Sleipner Project (Post-Combustion Capture)

The Sleipner project, located in the North Sea, is a pioneer in post-combustion carbon capture and storage (CCS). Operated by Equinor (formerly Statoil), this project captures carbon dioxide from the natural gas produced at the Sleipner West field. The captured carbon dioxide is then injected into a saline aquifer beneath the seabed for geological storage.

Achievements and Lessons Learned:

The Sleipner project has been operational since 1996 and has successfully captured and stored over 23 million tonnes of carbon dioxide.

It demonstrated that CCS is a viable option for reducing emissions from industrial sources.

The project's success has spurred the development of other CCS projects globally, fostering knowledge sharing and technological advancements.

2.3.2 SaskPower's Boundary Dam 3 (Pre-Combustion Capture)

The Boundary Dam Integrated Carbon Capture and Storage Project, located in Saskatchewan, Canada, is a pre-combustion carbon capture initiative. SaskPower retrofitted Unit 3 of the Boundary Dam Power

Station with carbon capture technology, capturing carbon dioxide before combustion.

Achievements and Lessons Learned:

The project has a capture capacity of around 1 million tonnes of carbon dioxide annually, making it one of the largest post-combustion capture facilities in the world.

The captured carbon dioxide is used for enhanced oil recovery in nearby oil fields.

The project demonstrates the feasibility of capturing carbon dioxide from existing power plants and utilizing the captured gas for economic benefit.

2.3.3 Climeworks (Direct Air Capture)

Climeworks, a Swiss company, specializes in direct air capture technology. Their modular carbon capture plants use sorbent materials to capture carbon dioxide directly from ambient air.

Achievements and Lessons Learned:

Climeworks' technology has been deployed in several locations, including Switzerland and Iceland.

The company's facilities can capture carbon dioxide for utilization in various applications, such as carbonation for beverages or for producing synthetic fuels.

Climeworks showcases the potential of direct air capture in removing carbon dioxide from the atmosphere and offers solutions for carbon utilization.

2.3.4 Petra Nova (Post-Combustion Capture)

The Petra Nova project, located in Texas, USA, is one of the world's largest post-combustion carbon capture facilities. It is integrated with the W.A. Parish power plant, a coal-fired facility. The project captures carbon dioxide from the flue gases before they are emitted into the atmosphere.

Achievements and Lessons Learned:

Petra Nova has a capture capacity of around 1.6 million tonnes of carbon dioxide annually.

The captured carbon dioxide is utilized for enhanced oil recovery, providing economic benefits while storing the gas underground.

The project demonstrates the potential for retrofitting existing fossil fuel power plants with carbon capture technology to reduce emissions.

2.3.5 Orca (Direct Air Capture)

The Orca direct air capture project, located in Iceland, is a joint effort by Carbfix, Climeworks, and ON Power. The project aims to capture carbon dioxide from ambient air and store it underground in basalt formations, where it mineralizes over time.

Achievements and Lessons Learned:

The Orca project is a pioneering example of combining direct air capture with mineralization as a form of carbon storage.

By injecting captured carbon dioxide into basalt rock, the gas reacts with minerals to form stable and inert compounds, providing a secure and long-lasting storage solution.

The project advances the concept of carbon removal and storage through natural geological processes.

2.3.6 Gorgon Carbon Dioxide Injection Project (CCUS)

The Gorgon Gas Project, located in Western Australia, includes a carbon capture, utilization, and storage (CCUS) component. It captures carbon dioxide from natural gas produced at the facility and injects it into a deep geological formation beneath Barrow Island.

Achievements and Lessons Learned:

The Gorgon project is one of the world's largest carbon capture and storage projects, capturing approximately 3.4 million tonnes of carbon dioxide annually.

It demonstrates the feasibility of CCUS in the natural gas industry and showcases the potential for reducing emissions from fossil fuel extraction.

2.3.7 Bioenergy with Carbon Capture and Storage (BECCS) Projects

BECCS projects combine bioenergy production with carbon capture and storage. These projects use biomass as a fuel source, capturing the carbon dioxide emitted during combustion and storing it underground.

Achievements and Lessons Learned:

BECCS projects contribute to negative emissions by removing more carbon dioxide from the atmosphere than they emit during bioenergy combustion.

The concept holds promise for achieving carbon neutrality or even carbon negativity, provided sustainable and environmentally sound biomass sourcing is ensured.

BECCS projects demonstrate the potential for synergizing energy production with carbon removal.

Conclusion

Chapter 2.3 presents case studies that highlight the successful implementation of different carbon capture technologies. These examples demonstrate the adaptability of carbon capture methods across various industries and geographies. By showcasing the achievements, challenges, and lessons learned from these projects, readers gain insights into the practical application of carbon capture technologies and their role in mitigating carbon emissions on a larger scale.

2.4 Assessment of the Scalability and Feasibility of Various Capture Approaches

As carbon capture technologies play a crucial role in addressing climate change, evaluating their scalability and feasibility is essential to determine their potential impact on a global scale. This section provides an in-depth assessment of the scalability and feasibility of different carbon capture approaches: post-combustion, pre-combustion, and direct air capture.

2.4.1 Post-Combustion Carbon Capture

Scalability: Post-combustion capture is well-suited for retrofitting existing industrial facilities and power plants. It offers a relatively straightforward integration process and can be applied to a wide range of sources, including coal-fired power plants and industrial processes. However, the scalability might be constrained by the availability of suitable solvents, as larger-scale implementations require significant quantities of these materials.

Feasibility: Post-combustion capture has been successfully demonstrated at several pilot and commercial-scale facilities. Its feasibility is evident through projects like the Sleipner project and Petra Nova, which have proven the technical viability and economic viability of capturing carbon dioxide from flue gases.

2.4.2 Pre-Combustion Carbon Capture

Scalability: Pre-combustion capture is well-suited for newly constructed facilities or those undergoing significant modifications. It can be implemented in integrated gasification combined cycle (IGCC) power plants, refineries, and other industrial installations that use syngas. While its scalability might be constrained by the availability of carbon-rich feedstocks and the need for gasification infrastructure, it has the potential to be applied across various sectors.

Feasibility: Projects like SaskPower's Boundary Dam 3 have demonstrated the feasibility of pre-combustion capture in a power plant setting. The project's success emphasizes its potential as a carbon mitigation strategy that can be integrated into industrial operations with the necessary infrastructure and resources.

2.4.3 Direct Air Capture

Scalability: Direct air capture has the advantage of being applicable anywhere, irrespective of the emission source. This versatility makes it highly scalable, with the potential to deploy units across various locations. However, the scalability might be constrained by factors such as the availability of suitable sorbents or solvents, the energy requirements for capturing carbon dioxide from dilute ambient air, and the challenges associated with scaling up operations.

Feasibility: Climeworks and other direct air capture projects have showcased the feasibility of capturing carbon dioxide directly from the atmosphere. While challenges remain, such as energy efficiency and cost-effectiveness, ongoing research and development are driving advancements in the feasibility of this approach.

Conclusion

Chapter 2.4 assesses the scalability and feasibility of different carbon capture approaches. Each approach has its strengths and challenges, and their potential for large-scale deployment depends on various factors including technological advancements, infrastructure availability, regulatory support, and economic considerations. Understanding the scalability and feasibility of these approaches is crucial for making informed decisions about which methods to prioritize and invest in to effectively mitigate carbon emissions and combat climate change.

CHAPTER THREE

Innovations in Carbon Storage and Utilization

3.1 Overview of Carbon Storage Options

The effective storage of captured carbon dioxide is a critical component of carbon capture and storage (CCS) strategies. This chapter explores various innovative carbon storage options, including geological storage, ocean storage, and mineralization.

3.1.1 Geological Storage

Geological storage involves injecting captured carbon dioxide into deep geological formations, such as depleted oil and gas reservoirs or deep saline aquifers. The principle behind geological storage is that the porous rock formations can trap and contain the carbon dioxide over long periods. This method leverages the natural trapping mechanisms that have held oil, gas, and brine in place for millions of years.

Benefits and Considerations:

Long-Term Storage: Geological formations can provide secure storage for carbon dioxide over geological timescales.

Proven Technology: The oil and gas industry has decades of experience with similar techniques, enhancing confidence in the feasibility of geological storage.

Monitoring and Verification: Careful monitoring and verification are required to ensure the stored carbon dioxide remains safely contained.

3.1.2 Ocean Storage

Ocean storage involves injecting carbon dioxide into the deep ocean, where it can dissolve in the seawater.

The dissolved carbon dioxide is transported and mixed into the vast oceanic volumes over time.

Benefits and Considerations:

Large Carbon Sink: The oceans are a vast carbon sink, capable of absorbing substantial amounts of carbon dioxide.

Natural Processes: The carbon dioxide can undergo various chemical reactions in the ocean, contributing to a complex carbon cycle.

Environmental Impact: Ocean storage poses potential risks to marine ecosystems and may alter ocean chemistry, affecting marine life.

3.1.3 Mineralization

Mineralization, also known as carbon mineralization or mineral carbonation, involves converting carbon dioxide into stable mineral forms through chemical reactions. This process mimics natural geological processes but is accelerated through human intervention.

Benefits and Considerations:

Permanent Storage: Mineralized carbon dioxide remains stable and immobilized in solid mineral structures.

End-Use Potential: Some mineral products formed through this process can have commercial applications, such as construction materials.

Energy and Resource Intensive: The mineralization process may require significant energy inputs and mineral resources.

Conclusion

Chapter 3.1 provides an overview of innovative carbon storage options, including geological storage, ocean storage, and mineralization. Each method presents unique opportunities and challenges in terms of safety, effectiveness, and environmental impact. By understanding the science and technology behind these options, societies can explore diverse avenues for safely storing captured carbon dioxide and reducing its atmospheric concentration.

3.2 Discussion of Emerging Technologies for Converting Captured Carbon into Useful Products (Carbon Utilization)

The concept of carbon utilization involves transforming captured carbon dioxide into valuable products, thereby turning a waste stream into a resource. This chapter explores a range of emerging technologies that hold the potential to convert captured carbon into useful products, contributing to economic value and environmental sustainability.

3.2.1 Carbon Capture and Utilization (CCU) Overview

Carbon capture and utilization (CCU) involves harnessing carbon dioxide as a feedstock to produce a variety of products, from chemicals and fuels to building materials. This approach not only reduces carbon emissions but also creates economic incentives for capturing and utilizing carbon dioxide.

3.2.2 Carbon-to-Chemicals and Fuels

Methanol Production: Carbon dioxide can be converted into methanol, a versatile chemical used in various industrial processes and as a fuel additive.

Synthetic Hydrocarbons: By combining captured carbon dioxide with hydrogen, synthetic hydrocarbons like synthetic natural gas or jet fuel can be produced.

Chemical Feedstocks: Carbon dioxide can serve as a feedstock for the production of various chemicals, including formic acid, ethylene, and more.

3.2.3 Carbon-to-Building Materials

Mineralization: Carbon dioxide can be reacted with minerals to form stable and durable building materials, such as aggregates and concrete.

Carbon-Fiber Composites: Carbon dioxide can be incorporated into the production of carbon-fiber composites, used in lightweight materials for aerospace and automotive industries.

3.2.4 Carbon-to-Value-Added Products

Bioplastics: Carbon dioxide can be utilized in the production of bioplastics, reducing the reliance on fossil fuel-based plastics.

Carbon-Based Materials: Carbon dioxide can be transformed into carbon-based materials, such as graphene and carbon nanotubes, with applications in electronics and materials science.

3.2.5 Challenges and Considerations

Economic Viability: The cost-effectiveness of carbon utilization technologies is a critical factor in their adoption and scalability.

Energy Inputs: Many carbon utilization processes require energy inputs, which should ideally come from renewable sources to ensure a net reduction in carbon emissions.

Regulatory and Market Factors: The regulatory environment and market demand for carbon-utilized products play a role in driving technology adoption.

Conclusion

Chapter 3.2 delves into the realm of emerging technologies for converting captured carbon dioxide into valuable products, thereby fostering carbon utilization. This approach holds significant promise for not only mitigating carbon emissions but also driving innovation, creating economic value, and addressing global sustainability challenges. As research and development continue, carbon utilization may revolutionize industries and contribute to a more circular and sustainable economy.

3.3 Exploration of Potential Industries and Applications for Carbon-Derived Materials

The versatility of carbon-derived materials offers a wide array of applications across various industries. This chapter explores the potential industries and applications where these materials can make a significant impact, contributing to sustainable practices and innovative solutions.

3.3.1 Construction and Building Materials

Concrete and Aggregates: Carbon-derived materials can be incorporated into concrete production, enhancing its strength, durability, and carbon sequestration potential.

Insulation Materials: Carbon-based materials can be used for creating efficient insulation, reducing energy consumption in buildings.

3.3.2 Energy Storage and Conversion

Batteries and Supercapacitors: Carbon-based materials can be used in electrodes for energy storage devices, improving their performance and lifespan.

Fuel Cells: Carbon-derived materials can serve as catalysts in fuel cells, enhancing their efficiency for clean energy conversion.

3.3.3 Electronics and Technology

Carbon Nanotubes and Graphene: Carbon-derived materials like carbon nanotubes and graphene have exceptional electrical and thermal conductivity, making them suitable for electronics, sensors, and more.

Transparent Conductive Films: Carbon-derived materials can replace traditional indium tin oxide for transparent conductive films used in displays and touchscreens.

3.3.4 Transportation and Automotive

Lightweight Materials: Carbon-derived materials can contribute to the development of lightweight and fuel-efficient materials for vehicles, reducing energy consumption.

Carbon Fiber Composites: Carbon fiber composites can be used to manufacture lighter and stronger automotive parts.

3.3.5 Agriculture and Food Production

Biostimulants: Carbon-derived materials can enhance soil fertility and plant growth as biostimulants in agriculture.

Carbon-Based Fertilizers: These materials can be incorporated into slow-release fertilizers for sustainable nutrient management.

3.3.6 Consumer Goods and Packaging

Biodegradable Plastics: Carbon-derived materials can be used to produce biodegradable plastics, reducing the environmental impact of single-use plastics.

Textiles and Clothing: Carbon-derived materials can enhance the performance of textiles, offering benefits like moisture-wicking and UV protection.

3.3.7 Pharmaceuticals and Medical Devices

Drug Delivery: Carbon-derived materials can be utilized as carriers for drug delivery systems, improving drug efficacy and minimizing side effects.

Biosensors: These materials can be incorporated into biosensors for medical diagnostics and monitoring.

3.3.8 Environmental Remediation

Water Purification: Carbon-derived materials can be used as adsorbents to remove pollutants from water, contributing to clean water supplies.

Air Filtration: These materials can be integrated into air filters to capture pollutants and improve indoor air quality.

Conclusion

Chapter 3.3 explores the diverse industries and applications that can benefit from carbon-derived materials. The potential to replace conventional materials with sustainable and innovative alternatives showcases the transformative role that carbon utilization can play in enhancing multiple sectors while contributing to a more sustainable and resource-efficient future.

3.4 Examination of Challenges and Opportunities in Implementing Carbon Storage and Utilization Projects

While carbon storage and utilization projects hold immense potential, they also face various challenges and offer opportunities that need to be carefully considered. This chapter delves into the key challenges and opportunities associated with implementing carbon storage and utilization projects.

3.4.1 Challenges

Economic Viability: Many carbon storage and utilization technologies require significant investments in infrastructure, research, and development. The cost-effectiveness of these projects can be a major challenge, particularly when compared to traditional practices.

Technology Readiness: Some carbon utilization technologies are still in the experimental or developmental stages. Scaling up these technologies to commercial levels while maintaining efficiency and performance can be challenging.

Regulatory Framework: The implementation of carbon storage and utilization projects often requires navigating complex regulatory landscapes. Ensuring compliance with environmental regulations, carbon pricing mechanisms, and permitting processes can be time-consuming and costly.

Energy and Resource Intensity: Some carbon utilization processes require substantial energy inputs, potentially offsetting the emissions reductions achieved. Additionally, the availability of key resources (e.g., suitable minerals for mineralization) can be limited.

Market Demand: The success of carbon utilization projects is linked to market demand for the resulting products. The market for carbon-derived materials and products may need time to develop and grow.

3.4.2 Opportunities

Emission Reductions: Carbon storage and utilization projects have the potential to significantly reduce carbon dioxide emissions, contributing to climate change mitigation goals.

Economic Growth: These projects can stimulate economic growth by creating new industries, jobs, and markets for carbon-derived products.

Technology Innovation: Developing and deploying carbon storage and utilization technologies encourages innovation and the advancement of sustainable solutions.

Circular Economy: Carbon utilization aligns with the principles of the circular economy, where waste streams are transformed into valuable resources, minimizing waste and promoting sustainability.

Carbon Removal: Some carbon utilization methods, such as mineralization, can result in the permanent removal of carbon dioxide from the atmosphere, contributing to negative emissions.

3.4.3 Collaboration and Knowledge Sharing

Industry Collaboration: Collaboration among industries, governments, research institutions, and NGOs is crucial for overcoming challenges and maximizing the potential of carbon storage and utilization projects.

Research and Development: Continued research and development are essential for improving the efficiency, scalability, and cost-effectiveness of carbon utilization technologies.

Knowledge Dissemination: Sharing lessons learned, best practices, and success stories can accelerate the adoption and deployment of these technologies globally.

Conclusion

Chapter 3.4 provides a comprehensive examination of the challenges and opportunities inherent in implementing carbon storage and utilization projects. While these projects face obstacles, their potential benefits, such as emissions reduction, economic growth, and innovation, underscore the importance of continued investment, collaboration, and technological advancement in the pursuit of sustainable solutions to address climate change and environmental concerns.

CHAPTER FOUR

Technological Advances and Breakthroughs

4.1 Survey of Recent Advancements in Carbon Capture Technology

This chapter explores the latest technological advancements and breakthroughs in carbon capture technology, focusing on novel materials, innovative processes, and cutting-edge research that are shaping the landscape of carbon capture and utilization.

4.1.1 Advanced Solvents and Sorbents

Ionic Liquids: Ionic liquids are emerging as efficient solvents for carbon capture due to their tunable properties and low volatility. Their unique chemical structure allows for enhanced carbon dioxide absorption and release, contributing to improved capture efficiency.

Metal-Organic Frameworks (MOFs): MOFs are porous materials with a high surface area, enabling them to adsorb large quantities of carbon dioxide. Researchers are designing MOFs with tailored properties for selective and efficient carbon capture.

4.1.2 Membrane-Based Separation Technologies

Polymeric Membranes: Advances in polymer chemistry have led to the development of highly selective membranes that allow for the separation of carbon dioxide from gas streams, offering energy-efficient alternatives to traditional absorption processes.

Mixed-Matrix Membranes (MMMs): MMMs combine polymers with nanoparticles or other fillers to enhance their separation performance, offering improved selectivity and permeability for carbon capture.

4.1.3 Direct Air Capture Innovations

Enhanced Sorbents: Researchers are designing and testing novel sorbent materials that can capture carbon dioxide from ambient air more effectively, offering higher capacity and faster kinetics.

Hybrid Approaches: Combining direct air capture with utilization pathways, such as mineralization or fuel production, can offer synergistic benefits and enhance the economic viability of the process.

4.1.4 Electrochemical Carbon Capture

Electrochemical Reduction: Electrochemical methods are being explored for capturing carbon dioxide by converting it into valuable chemicals and fuels, such as formate, methane, or ethylene, using renewable electricity.

4.1.5 Catalytic Conversion

CO2 Hydrogenation: Catalytic processes that convert carbon dioxide and hydrogen into valuable chemicals like methanol, formic acid, or even synthetic fuels are advancing with improved catalyst design and reaction pathways.

4.1.6 Artificial Photosynthesis

Solar-Driven Carbon Capture: Researchers are exploring artificial photosynthesis processes that mimic natural photosynthesis to capture carbon dioxide and convert it into useful products using sunlight.

4.1.7 Emerging Research and Collaboration

Interdisciplinary Research: The convergence of various scientific disciplines, such as chemistry, materials science, and engineering, is driving innovative solutions for carbon capture and utilization.

Global Collaboration: International collaboration and partnerships are accelerating research, allowing for knowledge exchange and sharing best practices to advance carbon capture technology on a global scale.

Conclusion

Chapter 4.1 presents a comprehensive survey of recent advancements in carbon capture technology, showcasing the innovative materials, processes, and research that are reshaping the landscape of carbon capture and utilization. These breakthroughs hold the promise of making carbon capture more efficient, cost-effective, and scalable, contributing to the broader goal of mitigating carbon emissions and addressing the challenges of climate change.

4.2 Showcase of Research and Development Efforts Aimed at Enhancing Efficiency and Cost-Effectiveness

This chapter highlights specific research and development efforts that are focused on advancing the efficiency and cost-effectiveness of carbon capture technologies. These initiatives demonstrate the dedication of researchers, engineers, and innovators to overcome challenges and drive progress in the field of carbon capture and utilization.

4.2.1 NET Power's Allam Cycle

The NET Power's Allam Cycle is an innovative natural gas power generation technology that integrates carbon capture directly into the combustion process. It uses a novel oxy-fuel combustion process with supercritical carbon dioxide (sCO_2) as the working fluid. This cycle enables efficient carbon dioxide capture without the need for additional separation steps, resulting in a near-zero-emission power generation process. The technology has the potential to produce low-cost electricity while capturing carbon dioxide for various utilization or storage options.

4.2.2 Carbon Clean Solutions' Modular Carbon Capture

Carbon Clean Solutions, a UK-based company, is focusing on developing modular carbon capture systems that can be retrofitted onto existing industrial processes. These systems utilize proprietary solvents and process designs to enhance the efficiency of carbon dioxide capture. By designing compact and scalable units, Carbon Clean Solutions aims to reduce the capital and operational costs associated with carbon capture technology deployment.

4.2.3 Project Tundra

Project Tundra is an initiative led by the Energy & Environmental Research Center (EERC) in North Dakota, USA. The project aims to retrofit an existing lignite coal-fired power plant with carbon capture technology. By incorporating advanced solvents and process optimization, Project Tundra seeks to demonstrate the feasibility of capturing carbon dioxide from a coal-fired facility while minimizing the associated costs and energy penalties.

4.2.4 Carbon Engineering's Direct Air Capture

Carbon Engineering, a Canadian company, is advancing direct air capture technology through engineering and cost optimization. Their Direct Air Capture facility in British Columbia is designed to capture carbon dioxide directly from ambient air. The company is continually refining the process to improve energy efficiency and reduce costs, aiming to demonstrate the viability of large-scale direct air capture as a means of removing carbon dioxide from the atmosphere.

4.2.5 Climeworks' DAC Commercialization

Climeworks, a Swiss company, is actively commercializing direct air capture technology. They are scaling up their modular carbon capture plants and exploring partnerships for various utilization pathways, including carbonation for beverages and producing synthetic fuels. Climeworks' efforts demonstrate how research and development, along with strategic collaborations, can contribute to making direct air capture a feasible and cost-effective solution.

4.2.6 Collaborative Research Initiatives

Numerous research collaborations between academic institutions, industries, and governments are driving technological advancements. Initiatives such as the Carbon Capture Coalition in the United States and international collaborations under organizations like the Global Carbon Capture and Storage Institute (GCCSI) are pooling resources and expertise to accelerate research, development, and deployment efforts.

Conclusion

Chapter 4.2 showcases ongoing research and development initiatives that are dedicated to enhancing the efficiency and cost-effectiveness of carbon capture technologies. These efforts underscore the commitment of stakeholders to address the challenges associated with carbon capture and utilization, ultimately contributing to the realization of sustainable and economically viable solutions for mitigating carbon emissions and combating climate change.

4.3 Analysis of Cross-Disciplinary Collaborations and Partnerships Driving Innovation in Carbon Capture

This chapter examines the significance of cross-disciplinary collaborations and partnerships in catalyzing innovation and driving advancements in the field of carbon capture and utilization. These collaborative efforts leverage diverse expertise, resources, and perspectives to address complex challenges and develop comprehensive solutions.

4.3.1 Collaboration Between Academia and Industry

Knowledge Exchange: Collaborations between academic researchers and industry professionals facilitate the exchange of scientific insights, technological know-how, and practical experience.

Innovation Acceleration: Industry partners provide real-world challenges that motivate researchers to develop innovative solutions with tangible applications, while academic expertise contributes scientific rigor and fresh perspectives.

4.3.2 Public-Private Partnerships

Resource Pooling: Public-private partnerships leverage government funding, industry investment, and research institutions' capabilities to advance carbon capture technology research and development.

Risk Mitigation: Government support can reduce the financial risk associated with developing new technologies, encouraging private-sector involvement in high-risk, high-reward ventures.

4.3.3 International Collaboration

Shared Knowledge: Collaborative efforts across countries and regions promote the sharing of best practices, lessons learned, and technological advancements.

Scale and Impact: Large-scale projects involving multiple countries can drive significant progress by pooling resources, expertise, and financial investments.

4.3.4 Multidisciplinary Approaches

Chemistry and Materials Science: Collaborations between chemists and materials scientists lead to the development of advanced sorbents, solvents, and catalysts that enhance carbon capture efficiency and selectivity.

Engineering and Process Optimization: Engineers work with scientists to design and optimize carbon capture processes, ensuring feasibility, scalability, and cost-effectiveness.

Economics and Policy: Collaboration between technical experts and economists enables the assessment of the economic viability and policy implications of carbon capture technologies.

4.3.5 Private Sector Collaboration

Industry Alliances: Private sector companies form alliances and consortiums to pool resources, share knowledge, and collectively drive technological advancements.

Value Chain Integration: Collaboration across the value chain, from carbon capture technology developers to end-users, ensures that solutions are holistic and aligned with market demands.

4.3.6 Knowledge Transfer and Education

Workforce Development: Collaborations between educational institutions, industry, and research organizations foster the development of skilled professionals and experts in carbon capture technology.

Dissemination of Research: Cross-disciplinary collaborations promote the publication and dissemination of research findings, contributing to the global knowledge base.

Conclusion

Chapter 4.3 highlights the pivotal role of cross-disciplinary collaborations and partnerships in propelling innovation in carbon capture and utilization. By uniting diverse expertise, these collaborations drive synergistic efforts that accelerate the development and deployment of carbon capture technologies, contributing to sustainable solutions for mitigating carbon emissions and addressing the challenges of climate change.

4.4 Exploration of Cutting-Edge Pilot Projects and Their Potential to Revolutionize the Field

This chapter delves into cutting-edge pilot projects that are pushing the boundaries of carbon capture and utilization technology. These projects showcase innovative approaches, demonstrate feasibility, and have the potential to revolutionize the field by paving the way for larger-scale implementation and widespread adoption.

4.4.1 Carbon Clean Solutions' Tuticorin Plant

Carbon Clean Solutions, in collaboration with Tuticorin Alkali Chemicals and Fertilizers Limited in India, has established a pilot project for capturing carbon dioxide emissions from a coal-fired power plant. The project utilizes Carbon Clean Solutions' proprietary solvent technology to capture carbon dioxide from flue gases. This pilot demonstrates the feasibility of retrofitting existing industrial facilities for carbon capture, laying the foundation for large-scale adoption in power generation and other industries.

4.4.2 Climeworks' Carbon Removal Plant

Climeworks' direct air capture plant in Iceland is a pioneering venture in removing carbon dioxide from the atmosphere on a commercial scale. The plant captures carbon dioxide directly from ambient air using specialized sorbent materials. The captured carbon dioxide is then injected into basaltic rock formations for mineralization. This pilot project showcases the potential of direct air capture to contribute to carbon removal and climate change mitigation, opening avenues for large-scale deployment and collaboration across industries.

4.4.3 Project Vesta's Enhanced Weathering

Project Vesta explores the concept of enhanced weathering, where olivine minerals are finely ground and spread on beaches, accelerating a natural process that absorbs carbon dioxide from the atmosphere through chemical reactions. This pilot project aims to test the effectiveness of this approach in sequestering carbon dioxide at a meaningful scale. If successful, enhanced weathering could revolutionize carbon removal strategies by leveraging natural geological processes to mitigate climate change.

4.4.4 Carbon Engineering's Direct Air Capture Facility

Carbon Engineering's direct air capture facility in Canada is a state-of-the-art project designed to capture carbon dioxide directly from ambient air. The facility utilizes large fans to draw in air, which is then passed through contactors containing specialized sorbents. The captured carbon dioxide is concentrated and can be used for various applications, including carbon utilization or storage. This pilot project demonstrates the technical feasibility of large-scale direct air capture, potentially transforming how carbon dioxide is managed in the atmosphere.

4.4.5 Advanced Clean Energy Storage Project

The Advanced Clean Energy Storage project in Texas, USA, combines carbon capture with energy storage to create an integrated energy solution. The project captures carbon dioxide emissions from a natural gas power plant and uses them to produce hydrogen via steam methane reforming. The hydrogen is then stored and used to generate electricity during peak demand periods. This innovative pilot project not only captures carbon dioxide but also contributes to energy storage and grid stability, showcasing the potential for carbon capture to integrate with other energy technologies.

Conclusion

Chapter 4.4 explores cutting-edge pilot projects that are on the forefront of revolutionizing the field of carbon capture and utilization. These initiatives highlight the practical applications, scalability, and potential impact of innovative technologies, setting the stage for transformative changes in how carbon emissions are managed, mitigated, and utilized to address the challenges of climate change.

CHAPTER FIVE

Pathways to Carbon-Zero Societies

5.1 Exploration of Policy Frameworks and Regulations

Promoting Carbon Capture and Reduction Efforts

This chapter delves into the critical role of policy frameworks and regulations in driving carbon capture and reduction efforts. It examines various strategies, incentives, and mechanisms that governments and international organizations are implementing to accelerate the transition towards carbon-zero societies.

5.1.1 Carbon Pricing and Market Mechanisms

Carbon Tax: Carbon taxes impose a fee on carbon emissions, incentivizing industries and individuals to reduce their carbon footprint by increasing the cost of emitting carbon dioxide.

Emissions Trading Systems (ETS): Emissions trading systems establish a cap on total emissions and allow companies to buy and sell emission allowances, encouraging efficient emission reduction while creating a market for carbon credits.

5.1.2 Renewable Energy Incentives

Renewable Energy Standards: Governments can mandate a certain percentage of energy production to come from renewable sources, stimulating the growth of low-carbon energy generation.

Feed-In Tariffs: Feed-in tariffs guarantee a fixed price for renewable energy producers, encouraging investment in renewable technologies and facilitating their integration into the energy mix.

5.1.3 Research and Innovation Funding

Government Grants: Governments allocate funds to research institutions and companies to develop and commercialize innovative carbon capture and utilization technologies.

Innovation Prizes: Prizes and competitions encourage entrepreneurs, scientists, and engineers to develop breakthrough solutions for carbon capture and reduction.

5.1.4 Regulatory Standards and Mandates

Emission Reduction Targets: Governments and international agreements set specific targets for reducing carbon emissions, motivating industries to adopt cleaner practices.

Energy Efficiency Standards: Regulations can require industries, buildings, and appliances to meet specific energy efficiency standards, reducing energy consumption and associated emissions.

5.1.5 Support for Carbon Capture and Utilization

Subsidies and Tax Credits: Governments provide financial

incentives to companies investing in carbon capture and utilization projects, reducing the upfront costs and encouraging adoption.

Low-Carbon Product Certifications: Certifying products with low carbon footprints promotes the demand for carbon-utilized materials and incentivizes manufacturers to adopt cleaner production processes.

5.1.6 International Collaboration and Agreements

Paris Agreement: International agreements like the Paris Agreement facilitate global cooperation in combating climate change, encouraging countries to enhance their carbon reduction efforts.

Technology Transfer: Collaborative projects between developed and developing countries promote the transfer of carbon capture technology and knowledge to regions with higher emissions.

5.1.7 Carbon Offsetting and Removal Programs

Reforestation and Afforestation: Programs that promote planting trees or converting barren land into forests serve as carbon sinks, offsetting emissions.

Carbon Removal Certificates: Creating a market for carbon removal certificates provides a financial incentive for projects that actively remove carbon dioxide from the atmosphere.

Conclusion

Chapter 5.1 explores the diverse policy frameworks and regulations that are driving carbon capture and reduction efforts globally. By creating incentives, setting targets, and promoting innovation, governments and international organizations play a pivotal role in shaping the pathways to carbon-zero societies, fostering sustainable practices and accelerating the transition towards a more environmentally responsible future.

5.2.1 Case Studies of Countries

Sweden: Sweden has made significant strides toward carbon neutrality by investing in renewable energy sources like hydroelectric, wind, and solar power. Their strong commitment to sustainability, energy efficiency, and green technologies has led to reduced carbon emissions while maintaining economic growth.

Bhutan: Bhutan is often considered a carbon-neutral country due to its focus on maintaining a carbon-negative balance. The country places a strong emphasis on forest conservation, renewable energy, and sustainable agricultural practices.

5.2 Case Studies of Cities

Copenhagen, Denmark: Copenhagen is known for its ambitious carbon-neutral goals. The city has invested in cycling infrastructure, efficient public transportation, and urban planning that prioritizes sustainability. These efforts have led to a significant reduction in carbon emissions and improved quality of life.

Reykjavik, Iceland: Reykjavik aims to become a carbon-neutral city by 2040. It relies on abundant geothermal energy for heating and electricity, reducing its dependence on fossil fuels. The city also promotes electric vehicles and encourages citizens to adopt sustainable practices.

5.3 Case Studies of Organizations

Microsoft: Microsoft is committed to becoming carbon-negative by 2030, aiming to remove more carbon from the atmosphere than it emits. The company is investing in renewable energy, carbon capture technologies, and afforestation projects to achieve this goal.

Interface: Interface, a carpet manufacturer, has embarked on a journey toward carbon neutrality by 2020. The company focuses on sustainable manufacturing processes, renewable energy use, and recycling materials. They aim to eliminate their carbon footprint while contributing to a circular economy.

5.4 Case Studies of Collaborative Efforts

The Climate Group's RE100 Initiative: The RE100 initiative brings together businesses committed to using 100% renewable energy. Companies like Google, Apple, and IKEA have joined this initiative, working collectively to reduce their carbon emissions and transition to clean energy sources.

Carbon Neutral Cities Alliance: This alliance consists of cities from around the world working together to share best practices and strategies for achieving carbon neutrality. Cities like Oslo, Vancouver, and Melbourne collaborate to accelerate their efforts and collectively address urban carbon emissions.

Conclusion

Chapter 5 highlights successful case studies of countries, cities, and organizations that have demonstrated their commitment to achieving carbon neutrality. These examples showcase a variety of approaches, strategies, and initiatives that have effectively reduced carbon emissions and paved the way toward a more sustainable and carbon-neutral future.

Public Perception and Awareness

Understanding Carbon Capture: Public awareness of carbon capture technologies varies widely. Many people may not fully understand the concept, potential benefits, and challenges associated with these technologies.

Media Influence: Media coverage can shape public perception, ranging from portraying carbon capture as a crucial solution to skepticism about its feasibility and effectiveness.

5.2 Factors Influencing Acceptance

Perceived Environmental Impact: Public acceptance can be influenced by concerns about the environmental impact of carbon capture technologies, including potential risks and unintended consequences.

Economic Considerations: People may be more accepting of carbon capture if they perceive economic benefits, such as job creation, innovation, and potential revenue streams from carbon utilization.

5.3 Education and Communication Strategies

Public Engagement: Governments, research institutions, and industries can engage the public through educational campaigns, workshops, and open discussions to improve understanding and address misconceptions.

Transparency: Openly sharing information about the benefits, challenges, and risks of carbon capture technologies can foster trust and informed decision-making.

5.4 Case Studies of Education Initiatives

Educational Workshops: Organizations like Carbon Clean Solutions conduct workshops to educate communities, students, and decision-makers about the science and potential of carbon capture technologies.

Interactive Exhibits: Museums and science centers create interactive exhibits that engage visitors in learning about carbon capture, making the technology more accessible and understandable.

5.5 Balancing Optimism and Realism

Communicating Progress: Highlighting successful pilot

projects, research advancements, and collaborations can inspire optimism and demonstrate that carbon capture is a viable solution.

Addressing Challenges: Acknowledging challenges and uncertainties while emphasizing ongoing research and commitment to improvement helps build realistic expectations.

5.6 Policy and Regulatory Impacts

Public Input: Involving the public in policy discussions and decision-making regarding carbon capture technologies can enhance acceptance and ensure a well-rounded approach.

Incentive Alignment: Effective policies that incentivize research, development, and deployment of carbon capture technologies can garner public support through their potential benefits.

Conclusion

This delves into the complex landscape of public perception, acceptance, and education regarding carbon capture technologies. By addressing misconceptions, fostering open communication, and engaging the public through education initiatives, societies can create a more informed and supportive environment for the development and implementation of carbon capture solutions.

Envisioning a Carbon-Zero Energy Future with Wide-Scale Carbon Capture Adoption

5.1 A Carbon-Zero Energy Landscape

Renewable Energy Dominance: In this envisioned future, renewable energy sources like solar, wind, hydro, and geothermal have become the primary energy sources, providing clean and abundant power to meet global energy demands.

Energy Storage Solutions: Advanced energy storage technologies enable the consistent supply of electricity from intermittent renewable sources, ensuring grid stability and reliability.

5.2 Integration of Carbon Capture Technologies

Industrial Emissions Mitigation: Carbon capture technologies are integrated into industrial processes across various sectors, effectively reducing carbon emissions from heavy industries like cement, steel, and chemicals.

Negative Emissions Balance: Wide-scale adoption of carbon capture, along with reforestation, enhanced weathering, and other carbon removal strategies, results in a net negative carbon emissions scenario, effectively combatting climate change.

5.3 Carbon Utilization and Circular Economy

Valuable Resource from Waste: Carbon capture technologies capture carbon dioxide emissions and transform them into valuable products, fostering a circular economy where waste becomes a resource.

Building Materials Revolution: Carbon-derived materials have revolutionized the construction industry, with carbon-neutral or carbon-negative building materials becoming the norm.

5.4 Technological and Economic Transformation

Innovation Hub: The rapid pace of innovation in carbon capture technologies has led to breakthroughs in efficiency, cost-effectiveness, and scalability, driving economic growth and job creation.

Global Collaboration: International partnerships have facilitated the exchange of knowledge, resources, and best practices, accelerating the global transition to carbon-zero energy systems.

5.5 Societal Mindset Shift

Sustainability as Norm: A fundamental shift in societal values places sustainability, environmental responsibility, and carbon neutrality at the forefront of decision-making, from personal choices to corporate strategies.

Education and Awareness: A well-informed public understands the importance of carbon capture and supports its integration into various aspects of daily life.

5.6 Policy and Regulatory Support

Forward-Thinking Policies: Governments worldwide have enacted progressive policies that incentivize carbon capture research, development, and implementation, further driving adoption and innovation.

Emission Reduction Targets: Aggressive emission reduction targets are achieved, and the focus shifts toward achieving net-negative emissions through carbon capture and utilization.

Conclusion

This Chapter presents a visionary outlook for a future powered by carbon-zero energy sources, where widespread carbon capture adoption has fundamentally transformed energy production, industries, and society as a whole. This future embodies a harmonious coexistence with the environment, marked by innovation, collaboration, and a collective commitment to addressing climate change through sustainable practices and technologies.

Epilogue: Paving The Path To A Sustainable Future

The journey through the chapters of "Carbon Zero: Harnessing Technology for Effective Carbon Capture" has explored the intricacies of carbon capture technologies, their scientific foundations, implementation strategies, and societal implications. As the narrative concludes, it is evident that the pursuit of carbon neutrality is not merely a scientific endeavor but a collective endeavor that transcends disciplines, borders, and generations.

The epilogue reflects on the insights gained from this exploration, emphasizing the urgency of addressing carbon emissions and the role that carbon capture plays in shaping a sustainable future. It encapsulates the key takeaways, from the importance of cross-disciplinary collaboration to the power of policy frameworks that support technological innovation.

Ultimately, the vision of a carbon-zero world, powered by renewable energy sources and supported by widespread carbon capture adoption, serves as a guiding light. It reminds us that the path to a sustainable future is both challenging and rewarding, requiring perseverance, innovation, and a shared commitment to safeguarding our planet for generations to come.

As we stand at the intersection of science, technology, policy, and society, the epilogue leaves readers with a call to action, inviting them to contribute to the realization of a carbon-zero world by embracing sustainable practices, advocating for change, and supporting the advancement of carbon capture technologies. With determination and collaboration, we can pave a path to a future where carbon is harnessed, not as a problem, but as a solution—a symbol of human ingenuity and our capacity to create a world that thrives in harmony with nature.

.....***.....

www.ingramcontent.com/pod-product-compliance
Lightning Source LLC
Chambersburg PA
CBHW071045290526
45795CB00004B/1335